Novare Chemistry Supplement

An introductory resource for students using
Chemistry for Accelerated Students

John D. Mays

Austin, Texas
2014

© 2014 John D. Mays

All rights reserved. Except as noted below, no part of this book may be reproduced or transmitted in any form or by any means, electronic or mechanical, including photocopying, recording, or by information storage and retrieval systems, without the written permission of the publisher, except by a reviewer who may quote brief passages in a review.

All images attributed to others under CC-BY-SA-3.0 or other Creative Commons licenses may be freely reproduced and distributed under the terms of those licenses.

Published by
Novare Science & Math
P. O. Box 92934
Austin, Texas 78709-2934
novarescienceandmath.com

Printed in the United States of America
ISBN: 978-0-9904397-0-7

For the complete catalog of textbooks and resources available from Novare Science and Math, visit novarescienceandmath.com.

Contents

Preface	iii
Chapter 1 Mass and Energy	1
Mass	1
Energy	2
The Law of Conservation of Energy	2
Forms of Energy	3
Chapter 2 Atoms and Atomic Theory	9
A Review of Atomic Facts	9
Ions	10
The History of Atomic Models	10
Chapter 3 Substances and Their Properties	16
Types of Substances	16
Pure Substances	16
Mixtures	19
Physical and Chemical Properties	20
Density	22
Image Credits	27

Preface

The purpose of the *Novare Chemistry Supplement* is to provide background to students who will be using our text *Chemistry for Accelerated Students* (CAS). CAS was specifically designed for accelerated sophomore students, concurrently enrolled in Algebra 2, who used our text *Accelerated Studies in Physics and Chemistry* (ASPC) the previous year as freshmen. As a result of studying ASPC the previous year, students have mastered the introductory principles of physics and chemistry and do not need to spend time at the beginning of chemistry (as most students do) reviewing basic principles from physical science and math. Structuring our texts this way—ASPC followed by CAS—makes the study of chemistry a completely different experience. Students have more time available in the course to cover standard material—an essential factor if the course is structured around a mastery paradigm, as CAS is. Additionally, mastery of important prerequisite math skills such as unit conversions and significant digits makes the computations in chemistry much more accessible, which removes a lot of the difficulty students often experience in a first-year chemistry course. Finally, mastery of key topics from physics enables students to *comprehend* the principles of chemistry, rather than simply memorizing. Together, these three benefits make the study of first-year chemistry uniquely accessible. Students learn more, experience less difficulty, and are less intimidated and stressed. Chemistry is a very challenging subject, but with the background from ASPC learning chemistry can be very rewarding.

Naturally, we would like for capable students to be able to use CAS for sophomore chemistry even if they did not study from ASPC the previous year. This supplement provides students with the necessary background to make that possible. Many of the topics in ASPC are revisited in CAS in a way that a student unfamiliar with ASPC can still readily follow. In other cases, prerequisite knowledge from ASPC is assumed and students without the necessary background will have difficulty following the explanation in CAS. These are the topics that are addressed in the *Novare Chemistry Supplement*. A perfect example is the topic of density. In CAS, the text assumes students know what density is, its units of measure, and how to calculate it. CAS does not explain those details but this supplement does.

Prior to starting the studies in this supplement, students who did not take ASPC should carefully study all of the topics in Appendix A of the main text, CAS. That appendix covers the metric system and metric prefixes, unit conversions, scientific notation, and the use of significant digits in problems requiring multiplication and division. (The addition rule is covered in the first chapter of CAS.) After going through Appendix A, students should work all of the exercises at the end. These exercises will give them practice at all the basic math skills listed above. This supplement includes additional practice using these skills in problems dealing with waves, work, and density.

Chapter 1   Mass and Energy

Matter, energy, and intelligence (or "information") are the three fundamental components the created world is made of.[1] Mass and energy concepts are ubiquitous in chemistry, so a firm grasp on these concepts is essential.

Mass

The best way to understand mass is to begin with *matter* and its properties. The term *matter* refers to anything composed of atoms or parts of atoms. Your thoughts, your soul, and your favorite song are not matter. You can write down your thoughts in ink, which *is* matter, and your song can be recorded onto a CD, which is matter. But ideas and souls are not material and are not made of what we call matter.

In this section we will focus on just two properties that all matter possesses: all matter takes up space and all matter has inertia. Describing and comparing these two properties will help make clear what we mean by the term *mass*.

All matter takes up space. Even individual atoms and protons inside of atoms take up space. Now, how do we *quantify* how much space an object takes up? That is, how do we put a numerical measurement to it? The answer is, of course, by specifying its *volume*. Volume is the name of the variable we use to quantify how much space an object takes up. When we say that the volume of an object is 338 cubic centimeters, what we mean is that if we could hollow the object out and fill it up with little cubes, each with a volume of one cubic centimeter, it would take 338 of them to fill up the hollowed object.

All matter has inertia. The effect of this property is that objects resist being accelerated. The more inertia an object has, the more difficult it is to accelerate the object. For example, if the inertia of an object is small, as with say, a golf ball, the object will be easy to accelerate. Golf balls are easy to throw, and if you hit one with a golf club it will accelerate at a high rate to a very high speed. But if the amount of inertia an object has is large, as with say, a grand piano, the object will be difficult to accelerate. Just try throwing a grand piano or hitting one with a golf club and you will see that it doesn't accelerate at all. This is because the piano has a great deal more inertia than a golf ball.

As with the property of taking up space, we need to be able to quantify an object's inertia. The way we do this is with the variable we call *mass*. The mass of an object is a numerical measurement specifying the amount of inertia the object has. Since inertia is a property of matter, and since all matter is composed of atoms, it should be pretty obvious that the more atoms there are packed into an object, the more mass it will have. And since the different types of atoms themselves have different masses, an object made of more massive atoms will have more mass than an object made of an equal number of less massive atoms.

The main unit of measure we use to specify an object's mass is the *kilogram*. There are other units such as the gram and the microgram. The kilogram (kg) is one of the base units in the SI system of units (the metric system). On the earth, an object weighing 2.2 pounds (lb) has a mass of one kilogram. To give you an idea of what a kilogram mass feels like in your hand on

---

[1] In this short chapter we will only discuss mass and energy. Those interested in a more detailed treatment of all three of these basics may wish to refer to my text, *Novare Physical Science*. See novarescienceandmath.com.

Chapter 1

Figure 1-1. This battery has a mass of 1 kg.

the earth, the lantern battery pictured in Figure 1-1 weighs 2.2 lb, and thus has a mass of 1 kg.

We have established that the mass of an object is a measure of its inertia, which in turn depends on how many atoms it is composed of and how massive those atoms are. The implication of this is that an object's mass does not depend on where it is. A golf ball on the earth has the same mass as a golf ball at the bottom of the ocean, on the moon, or in outer space. Even where there is no gravity, the mass of the golf ball will be the same. This is what distinguishes the *mass* of an object from its *weight*.

Weight is caused by the force of gravity acting on an object composed of matter (which we often simply refer to as *a mass*). The weight of an object depends on where it is. An object—or mass—on the moon only weighs about 1/6 of its weight on earth, and in outer space, where there is no gravity, a mass has no weight at all. But the mass of an object does not depend on where it is. This is because an object's mass is based on the matter the object is made of. The lantern battery in the figure has a certain weight on the earth (2.2 lb). In outer space, it weighs nothing and will float right in front of you. But if you try to throw the battery, the force you feel on your hand will be the same on the earth or in space. That's because the force you feel depends on the object's mass.

Here is a summary using slightly different terminology. Inertia is a *quality* of all matter; mass is the *quantity* of a specific portion of matter. Inertia is a quality or property all matter possesses. Mass is a quantitative variable, and it specifies an amount of matter, a quantity of matter.

Energy

Defining energy is tricky. Dictionaries usually say, "the capacity to do mechanical work," which is not particularly helpful. So we are not going to try to define it accurately, we are just going to accept that energy exists in the universe, it was put there by God, and it exists in many different forms. It is fairly obvious that a bullet traveling at 2,000 ft/s has more energy than a bullet at rest. This is why the high speed bullet can kill but the bullet at rest cannot.

The Law of Conservation of Energy

The *law of conservation of energy* states that energy can be neither created nor destroyed, only changed in form. Energy can be in many different forms in different types of substances, such as in the molecules of gasoline, in the waves of a beam of light, in heat radiating through space, in moving objects, in compressed springs, or in objects raised vertically on earth. As different physical processes occur such as digesting food, throwing a ball, operating a machine, heating due to friction, or accelerating a race car, energy in one form is being converted into some other form. Energy might be in one form in one place, such as in the chemical potential energy in the molecules in the muscles of your arm, and be converted through a process like throwing a ball to become energy in another form in another place, like kinetic energy in the ball.

Strictly speaking, the law of conservation of energy is violated if nuclear processes are considered. This is because of the so-called mass-energy equivalence discovered by Albert

Mass and Energy

Einstein and forever enshrined in his famous equation, $E = mc^2$. In this equation E represents energy, m represents mass, and c represents the speed of light. As you can see, the constant of proportionality between mass and energy is the speed of light squared, a very large number. This explains why nuclear reactions, such as take place in stars and in nuclear weapons, release such a huge amount of energy. A tiny quantity of mass multiplied by c^2 results in a lot of energy. During nuclear reactions mass is actually converted into energy, creating energy when none was there before. For this reason we now say that the law of conservation of energy is only considered to hold across the board if the energy equivalence of mass is included.

Forms of Energy

There are many different forms of energy, some of which are described below. The first one, gravitational potential energy, is important to know about because we humans have a lot of direct experience with the way objects behave in gravitational fields. This makes gravitational potential energy a useful model or analogy for describing another form of energy we are less familiar with—electrical potential energy.

- *Gravitational Potential Energy*

  In a gravitational field such as we have on Earth, energy is required to lift an object vertically. If a crane hoists an object above the ground, the object will then possess *gravitational potential energy*. The term *potential* in the name of this form of energy indicates that the energy is stored and has the capability of converting into another form of energy when released. After having been raised up by the crane, if the object is released it will fall, and the gravitational potential energy is has will begin converting into kinetic energy, the energy associated with an object's motion. Immediately before hitting the ground, all of the gravitational potential energy the object had will have been converted into kinetic energy. To calculate gravitational potential energy, E_G, we use the equation

  $E_G = mgh$

  In this equation, m is the object's mass, g is the acceleration due to gravity on earth (9.80 m/s²), and h is the object's height. Thus E_G is directly proportional to both an object's mass and its height.

- *Kinetic Energy*

  Kinetic energy is the energy an object possesses because it is in motion. Symbolized as E_K, kinetic energy is calculated as

  $E_K = \frac{1}{2} mv^2$

  The faster an object is moving, the more kinetic energy the object has. The kinetic energy of a moving object is directly proportional to its mass, and proportional to the square of the object's velocity.

- *Electromagnetic Radiation*

  Electromagnetic radiation is itself pure energy, and propagates in the form of electromagnetic waves traveling through space, or through media such as air or glass. This type of energy includes all forms of light (infrared, visible, ultraviolet, etc.), as well as radio waves, microwaves, x-rays, and gamma rays. Together, these make up the *electromagnetic spectrum*, a vast spectrum of electromagnetic radiation found in nature. The terms *light* and *electromagnetic radiation* are essentially synonyms, although in common speech we often use the term light to refer to light we can see with our eyes. But the other terms for electromagnetic radiation (microwaves, radio waves, x-rays, etc.) all refer to this same form of energy. The only difference is in the *wavelength* of the electromagnetic waves. Graphically, the wavelength of a wave may be represented as shown in Figure 1-2. The velocity of a wave relates to the wavelength and frequency of the wave as follows:

$$v = \lambda f$$

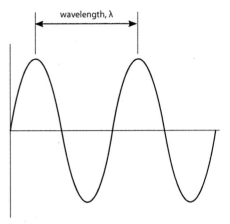

Figure 1-2. The parts of a wave. Two complete cycles of the wave are shown in the figure.

In this equation, v is the velocity of the wave, which for electromagnetic radiation is the speed of light measured, in units of meters per second (m/s). The other variables are λ ("lambda," the Greek letter l), the wavelength of the wave measured in meters (m), and f, the frequency of the wave measured in units called hertz (Hz). The *frequency* of a wave is the number of wave cycles the wave completes in one second. The unit "hertz" simply means cycles per second. So if a wave completes 5,000 cycles per second, its frequency is 5,000 Hz or 5 kHz.

As you can see from this equation, the velocity of a wave varies directly with its wavelength and directly with its frequency. But physically, it is better to think of the wavelength as depending on the other two variables in this equation. This is because the wave velocity actually depends on the medium in which the wave is propagating. Also, the frequency depends on the source of the wave; whatever is causing the wave is oscillating at a certain rate, and this rate determines the frequency of the wave being produced. So since the wave velocity is determined by the medium, and the frequency is determined by the source of the wave, this means the wavelength is the variable that really depends on the other two. So solving the wave equation for the wavelength as the dependent variable, we have

$$\lambda = \frac{v}{f}$$

From this form of the equation we see that the wavelength and frequency of a wave are inversely proportional. Higher frequencies have shorter wavelengths.

Here is an example calculation using this equation.

Radio station KUT FM in Austin broadcasts a carrier signal at 90.5 MHz. Determine the wavelength of this wave.

All radio waves (FM, AM, short wave, etc.) are part of the electromagnetic spectrum and propagate at the speed of light. We begin by writing the given information. Then we perform the needed unit conversions so that all quantities are in MKS (meter-kilogram-second) units.

$$f = 90.5 \text{ MHz} \cdot \frac{1 \times 10^6 \text{ Hz}}{1 \text{ MHz}} = 9.05 \times 10^7 \text{ Hz}$$

$$v = 2.998 \times 10^8 \text{ } \frac{\text{m}}{\text{s}}$$

Next, we write the equation, insert the values, and compute the result.

$$\lambda = \frac{v}{f} = \frac{2.998 \times 10^8 \text{ } \frac{\text{m}}{\text{s}}}{9.05 \times 10^7 \text{ Hz}} = 3.31 \text{ m}$$

The given frequency has three significant digits, so the result does as well.

- *Electrical Potential Energy*

The best way to understand electrical potential energy is by analogy with gravitational potential energy. Objects with mass exert a gravitational attraction on one another. We describe this phenomenon by saying that a mass generates a gravitational field around itself, and any other object entering the field will feel the attraction of the gravitational force. In the same way, objects with electrical charge exert an electrical attraction or repulsion on each other. We describe this effect by saying that a charged particle generates an electrical field around itself, and any other charged particle entering the field will feel the attraction or repulsion of the electrical force.

As explained in the Introduction to the main text, the concept of the electrical forces between charged particles (protons and electrons) is one of the central organizing ideas in chemistry. Interestingly, the mathematical descriptions of electrical and gravitational fields are structurally identical, and this is why gravitational fields and gravitational potential energy are so useful in helping us to understand electrical fields and electrical potential energy. In fact, for our purposes there are only two main differences between the gravitational and electrical forces. First, the electrical force is 10^{36} times as strong as the gravitational force! To give you some idea of how huge this number is, consider this. If you had a stack of 10^{36} sheets of copy paper, the height of this stack would be a distance that is about 100 billion (100,000,000,000) times the diameter of the entire Milky Way galaxy, which itself is 100,000 light years across. The bottom line is that in chemistry, gravity basically doesn't matter at all because the electrical forces between charged particles are so colossal by comparison. Second, gravitational forces are always attractive but electrical forces can be repulsive or attractive, depending on whether the signs of the charges involved are the same or opposite. Thus, gravitational potential energy appears when objects attracted to one another are pulled apart. The potential energy between oppositely charged particles works the same way. But particles possessing like charge (both positive or both negative) repel each other, so potential energy appears when two such charges are pushed together.

Chapter 1

Natural processes like chemical reactions tend to go in a direction that minimizes the potential energy. This is illustrated in Figure 1-3. Left to themselves, objects attracted by the gravitational force will come together (a), which is why objects fall to the ground when released. By falling together, the potential energy between them is minimized. Objects experiencing electrical forces minimize the potential energy between them either by coming together (b) or by pushing apart (c).

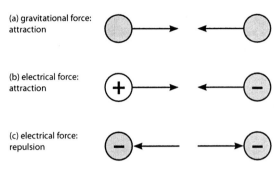

Figure 1-3. Comparison of gravitational and electrical forces.

- *Electrical Energy*

Since around 1800, we have known how to transport energy in electrical form in wires. These days just about everything is (or can be) powered electrically. For our study of chemistry, electrical energy will only play a role toward the end of the course when we study oxidation-reduction (redox) chemistry. In the redox chapter we will address the subject of electrochemistry, and at that point energy flowing in electrical wires will come into play.

Electricity is the flow of electrons in wires. Electrons are negatively charged, and in an electrical field electrons will flow toward the positive terminal of a battery or other power source. In a more general sense, any flow of charged particles constitutes an electric current. In copper wires, the moving charges are electrons. In an aqueous solution (a solution in water), the charges are often ions—atoms or clusters of atoms with a net charge.

- *Thermal Energy and Internal Energy*

Thermal energy is a general term that refers to the energy a hot object or substance possesses because it has been heated. The *internal energy* of a substance is the total of all of the kinetic energies possessed by the atoms or molecules of the substance. Atoms or molecules are constantly in motion, vibrating or translating, or both. Atoms in solids cannot fly around, so they vibrate in place, but atoms in fluids are free to translate, or move around. Either way, since atoms are always moving each one possesses an average kinetic energy, $E_K = \frac{1}{2}mv^2$. If you add up the kinetic energy of every particle in a certain substance, that total is its internal energy. This term is much more precise than the term thermal energy, and can actually be computed.

The internal energy of a substance correlates directly to its temperature. The higher the temperature, the higher the internal energy in the substance, which in turn implies faster moving particles. The concepts of internal energy and temperature relate immediately to the concept of absolute zero. Absolute zero is the zero temperature on the Kelvin temperature scale. The temperature of a substance varies directly with its internal energy. A temperature of 0 kelvins (absolute zero) means no internal energy at all, or in other words, atoms standing still! Atoms can't move any slower than standing still, and thus, there is no temperature lower than absolute zero. As far as we know, there is no place in the universe where the temperature is absolute zero, although physicists in low-temperature research labs have succeeded in achieving temperatures of only a few millionths of a degree above absolute zero.

- *Work*

    In physics, the term *work* denotes a mechanical process by which a specific amount of energy is transferred from one object to another. Objects do not possess work energy, as they do other with forms of energy. Instead, one object "does work" on another by applying a force to it and moving it a certain distance. When one object does work on another, energy is transferred from the first object to the second.

    The way an object acquires kinetic energy or gravitational potential energy is that another object or person or machine does work on it. Work is the way mechanical energy is transferred from one machine or object to another. Work is defined as the energy it takes to push an object with a certain (constant) force over a certain distance. Work is calculated as

    $$W = Fd$$

    where F is the force on an object, measured in newtons (N), d is the distance it moves (m), and W is the work done on the object, measured in joules (J), the standard unit of measure in the SI system for quantities of energy.

    If you push a person on a bicycle over a certain distance, you deliver energy to the person on the bicycle equal to the pushing force times the distance pushed. Assuming there is no friction, the work energy that comes from the pusher (you) is now in the kinetic energy of the person on the bicycle. (If there is friction, then some of the energy will convert into heat.) As another example, if you raise an object up above the ground you are doing work on the object. The force required to lift it is its weight, so the energy required to lift an object is the object's weight times the height it is lifted.

- *Heat*

    In physics, the term *heat* denotes energy in transit, flowing by various means from a hot substance to a cooler substance when a difference in temperature is present. Notice that heat is like work—both terms describe processes by which specific amounts of energy are transferred from one place to another. As with work, objects do not possess heat. After being heated, we would speak of an object's thermal energy or its internal energy.

    Heat is always absorbed or released in chemical reactions. Consider holding a beaker in your hand while a chemical reaction takes place inside the beaker. If the beaker feels hot, it is because the reaction is releasing energy in the form of heat, and you can feel this heat. Heat flowing into the water from the reaction in the beaker increases the internal energy in the water, and thus its temperature. Any process that releases heat like this is said to be *exothermic*. If the beaker feels cold, it is because the reaction is absorbing energy in the form of heat, and heat is flowing from the water in the beaker into the compounds being formed by the reaction. Since heat is flowing out of the H_2O and into the compounds formed by the reaction, the internal energy of the water decreases, along with its temperature.

    There are three different mechanisms by which heat can flow. The first is *conduction*, which occurs primarily in solids. In conduction, absorption of heat by the atoms in one part of the solid causes the atoms to vibrate more vigorously. This means higher kinetic energy and higher temperature. These vigorous vibrations then spread atom by atom throughout the solid. In metals, the spread of heat is also facilitated by the free electrons that collide and spread heat through the metal.

The second heat transfer mechanism is *convection*. Convection occurs in liquids and gases where molecules are free to fly around, collide, and mix and mingle. This process allows the thermal energy in a hot fluid (liquid or gas) to spread into a region of cooler fluid, and the kinetic energy is spread molecule to molecule through molecular collisions.

The third heat transfer mechanism is *electromagnetic radiation*. The region of the electromagnetic spectrum primarily associated with heat is the infrared region. When you feel warm standing in the sun or in front of a fire it is because the energy of infrared electromagnetic radiation is being absorbed by your skin. A fire is an exothermic chemical reaction, and fires release a lot of infrared electromagnetic radiation.

Questions

1. Distinguish between matter and mass.
2. Explain the difference between mass and weight.
3. In reference to waves, what does the term *frequency* refer to?
4. What makes work and heat different from the other forms of energy described in this chapter?
5. In what ways are electrical potential energy and gravitational potential energy similar?
6. What are two significant ways that the forces of electrical attraction and gravitational attraction are different?
7. How does the internal energy of a substance relate to the temperature of that substance?
8. When an object falls from a given height, it accelerates downward and its kinetic energy increases. According to the law of conservation of energy, energy is not created in this scenario; it comes from somewhere else. Explain where the increasing kinetic energy of a falling object is coming from. Then speculate on where the energy might have been before that.
9. Distinguish between thermal energy and internal energy.
10. Determine the wavelength of a radio signal broadcasting at 1,310 kHz. *(Ans: 229 m)*
11. In air, sound travels at approximately 342 m/s. The wavelength of the lowest note on a standard electric bass guitar is 8.30 m. Determine the frequency of the sound wave. *(Ans: 41.2 Hz)*
12. Determine the frequency of the light waves in a laser beam if the laser has a wavelength of 532 nm. State your result in GHz. *(Ans: 6.45×10^7 GHz)*
13. Determine the energy (work) required for a force of 2.11×10^{-3} N to move an object 9.50 m. *(Ans: 0.0200 J)*
14. Imagine that 175.0 mJ of energy are required to move an object 1,600.0 μm. What force was required to move the object? *(Ans: 109,400 N)*

Chapter 2   Atoms and Atomic Theory

A Review of Atomic Facts

All matter is made of atoms, the smallest basic units matter is composed of. An atom of a given element is the smallest unit of matter that possesses all of the properties of that element.

Atoms are almost entirely empty space. Each atom has an incredibly tiny nucleus in the center containing all of the atom's protons and neutrons. Since the protons and the neutrons are in the nucleus, they are collectively called *nucleons*. The masses of protons and neutrons are very nearly the same, although the neutron mass is slightly greater. Each proton and neutron has nearly 2,000 times the mass of an electron, so the nucleus of an atom contains practically all of the atom's mass. Outside the nucleus is a weird sort of cloud surrounding the nucleus containing the atom's electrons.

We will save the details about electrons for the main text, but here is a brief preview. The electron cloud consists of different *orbitals* where the electrons are contained. Electrons are sorted into the atomic orbitals according to the amount of energy they each have. For an electron to be in a specific orbital means the electron has a certain amount of energy—no more, no less.

We can say that atoms are almost entirely empty space because the nucleus is incredibly small compared to the overall size of the atom with its electron cloud. It's quite easy for us to pass over that remark without pausing to consider what it means. To help visualize the meaning, consider the athletic stadium picture in Figure 2-1. Using this stadium as an enlarged atomic model, the electrons in their orbitals would be zipping around in the region where the seating sections are in the stadium. Each electron in this vast space is far smaller than the period at the end of this sentence. The atomic nucleus containing the protons and neutrons is located at the center of the playing field and is the size of a pinhead. The rest of the space in the atom is completely empty. Nothing is there, not even air, since air, of course, is also made of atoms.

Returning to our discussion of atomic facts, one of the fundamental physical properties of the subatomic particles is *electric charge*. Neutrons have no electric charge. They are electrically neutral, hence their name. Protons and electrons each contain exactly the same amount of charge, but the charge on protons is positive and the charge on electrons is negative. If an atom or molecule has no net electric charge, it contains equal numbers of protons and electrons.

Atoms are significantly smaller than the wavelengths of light (about 5,000 times smaller), which means light does not reflect off atoms and there is no way to see them. The same is true of *molecules*. Molecules are clusters of atoms chemically bonded together. When atoms of different elements are bonded together in a mol-

Figure 2-1. The head of a pin at center field in a stadium is analogous to the nucleus in the center of an atom.

ecule they form a compound, which we will discuss in the next chapter of this booklet. But sometimes atoms of the same element bond together in molecules, as illustrated in Figure 2-2. Oxygen, chlorine, nitrogen, and hydrogen are common *diatomic gases* that form molecules consisting of a pair of atoms chemically bound together.

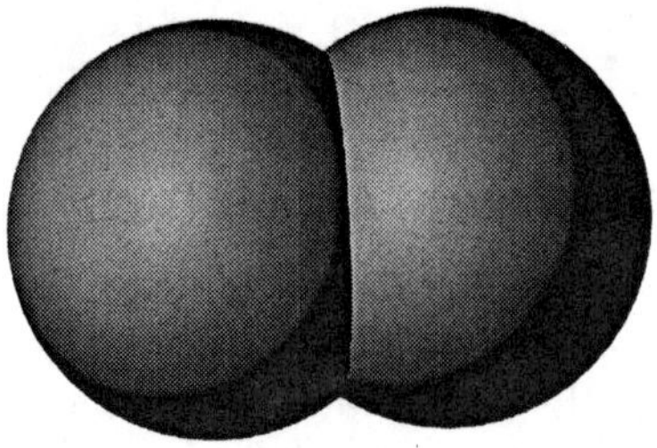

Ions

An atom that is electrically neutral possesses an equal number of protons and electrons. This is the way we normally think of isolated atoms. But in many chemical process, atoms gain or lose electrons and thus have a net charge equal to the number of electrons gained or lost. If the atom gains electrons, its net charge is negative. If the atom loses electrons, it then has more protons than electrons and its net charge is positive. Any atom that has a net charge is referred to as an *ion*. Ions come up in nearly every chapter of the main text.

Figure 2-2. Space-filling model of a diatomic molecule.

The process of gaining or losing electrons is called *ionization*. For reasons that will be explained in the main text, the ways atoms of some elements will ionize are very predictable. Have a look at the Periodic Table of the Elements you will find inside the back cover of your text (*Chemistry for Accelerated Students*). The elements you see listed in the first column—which is called Group 1—all ionize by losing one electron. All the elements listed in Group 2 ionize by losing two electrons. On the right end of the periodic table, elements in Group 18 don't ionize at all.[1] These elements are called the noble gases. Elements in Group 17 ionize by gaining one electron, and elements in Group 16 ionize by gaining two electrons. Similarly, most of the elements in Group 3 ionize to +3, and the first two elements in Group 15 ionize to −3.

If an atom of an element ionizes by gaining an electron, as chlorine does (Group 17), the atom is then an ion with a charge of −1. We denote this by placing the charge as a superscript on the element's chemical symbol, as in Cl^-. If an ion such as calcium has a charge of +2, (Group 2), then we write Ca^{2+}. Notice that if the ionic charge is +1 or −1, the 1 is not written in the superscript. Notice also that in the superscript on an ion we conventionally write the sign of the charge after the value of the charge.

Ions can also form where several atoms are bonded together in a molecule and as a group they have a net charge. These ions are called *polyatomic* ("many-atomed") *ions*. Common examples are hydroxide, OH^-, carbonate, CO_3^{2-}, and sulfate, SO_4^{2-}. Polyatomic ions are discussed in detail in the main text, so we will leave the rest of this topic for treatment there.

The History of Atomic Models

The story of atomic theory starts back with the ancient Greeks. As we look at how the contemporary model of the atom developed, we will hit on some of the great milestones in the history of chemistry and physics along the way.

---

[1] Except under extreme conditions, such as those inside stars, plasmas, or specially contrived in laboratories.

Atoms and Atomic Theory

In the 5th century BC, the Greek philosopher Democritus proposed that everything was made of tiny, indivisible particles. Our word atom comes from the Greek word *atomos*, meaning "indivisible." Democritus' idea was that the properties of substances were due to characteristics of the atoms they are made from. So atoms of metals were supposedly hard and strong, atoms of water were assumed to be wet and slippery, and so on. At this same time, there were various views about what the most basic substances—that is, the elements—were. One of the most common views was that there were four elements—earth, air, water, and fire—and that everything was composed of these.

Not much real chemistry went on for a very long time. During the medieval period, of course, there were the alchemists who sought to transform lead and other materials into gold. But this cannot be done by the methods available to them, so their efforts were not successful.

But in the 17th century things started changing as scientists became interested in experimental research. The goal of the scientists described here was to figure out what the fundamental constituents of matter were. This meant figuring out how atoms were put together, what the basic elements were, and understanding what was going on when various chemical reactions took place. The nature of earth, air, fire, and water was under intense scrutiny over the next 200 years.

In 1803, English scientist John Dalton (Figure 2-3) produced the first scientific model of the atom. Dalton's atomic model was based on five main points, listed in Table 2-1.

The impressive thing about Dalton's atomic theory is that even today the last three of these points are regarded as correct, and the first two are at least partially correct. On the first point, it is still scientifically factual that all substances are made of atoms, but we now know that atoms are not indivisible. This is now obvious, since atoms themselves are composed of protons, neutrons and electrons. The second point is correct in every respect but one. Except for the number of neutrons in the nucleus, every atom of the same element is identical. However, we now know that atoms of the same element can vary in the number of neutrons they have in the nucleus.

Figure 2-3. English scientist John Dalton.

After Dalton, the next breakthrough in our understanding of atomic structure came from English scientist J.J. Thomson (Figure 2-4). Thomson worked at the Cavendish Laboratory in Cambridge, England. In 1897, he conducted a series of landmark experiments that revealed the existence of electrons. Because of his work he won the Nobel Prize in Physics in 1906. A photograph of the *cathode ray tube* Thomson used for his work is shown in Figure 2-5.

Thomson placed electrodes from a high-voltage electrical source inside of a very elegantly made sealed-glass vacuum tube. This apparatus can generate a so-

1. All substances are composed of tiny, indivisible substances called atoms.
2. All atoms of the same substance are identical.
3. Atoms of different elements have different weights.
4. Atoms combine in whole-number ratios to form compounds.
5. Atoms are neither created nor destroyed in chemical reactions.

Table 2-1. The five tenets of Dalton's 1803 atomic model.

11

Chapter 2

Figure 2-4. English scientist J.J. Thomson.

called *cathode ray* from the negative electrode (1), called the cathode, to the positive one, called the anode (2). A cathode ray is simply a beam of electrons, but this was not known at the time. The anode inside Thomson's vacuum tube had a hole in it for some of the electrons to escape through, which created a beam of "cathode rays" heading toward the other end of the tube (5).

Thomson placed the electrodes of another voltage source inside the tube (3), above and below the cathode ray and discovered that the beam of electrons deflected when this voltage was turned on. He also placed magnetic coils on the sides of the tube (4) and discovered that the electrons also deflected as they passed through the magnetic field produced by the coils. The deflection of the beam toward the positive electrode led Thomson to theorize that the beam was composed of negatively charged particles, which he called "corpuscles." (The name *electron* was first used a few years later by a different scientist.) By trying out many different arrange-

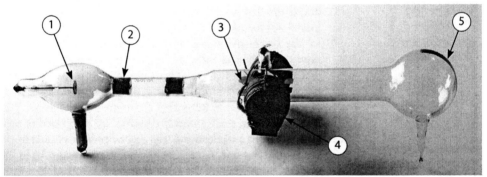

Figure 2-5. Thomson's cathode ray tube.

ments of cathode ray tubes, Thomson confirmed that the ray was negatively charged. Then using the scale on the end of the tube to measure the deflection angle (5), he was able to determine the charge-to-mass ratio of the individual electrons he had discovered, which is 1.8×10^{11} C/kg, where C stands for coulomb, the SI unit of electric charge.

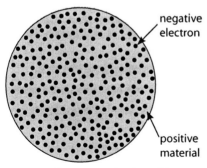

Figure 2-6. Thomson's plumb pudding model of the atom.

Thomson went on to theorize that electrons came from inside atoms. He developed a new atomic model that envisioned atoms as tiny clouds of massless, positive charge sprinkled with thousands of the negatively charged electrons, depicted in Figure 2-6. Thomson's model is usually called the *plum pudding model*.

In 1911, American scientist Robert Millikan

Atoms and Atomic Theory

Figure 2-7. American scientist Robert Millikan.

(Figure 2-7) devised his famous *oil-drop experiment*, an extraordinary procedure that allowed him to determine the charge on individual electrons, 1.6×10^{-19} C. Once this value was known, Millikan used Thomson's charge-to-mass ratio and calculated the mass of the electron, 9.1×10^{-31} kg. Millikan's apparatus is pictured in Figure 2-8.

Inside of a heavy metal drum (1) about the size of a 5-gallon bucket, Millikan placed a pair of horizontal metal plates connected to an adjustable high-voltage source. The upper plate had a hole in the center and was connected to the positive voltage, the lower plate to the negative. He used an atomizer spray pump (2) to spray in a fine mist of watchmaker's oil above the positive plate. Some of these droplets would fall through the hole in the upper plate and move into the region between the plates. Connected through the side of the drum between the two plates was a telescope eyepiece (3) and lamp so that Millikan could see the oil droplets between the plates.

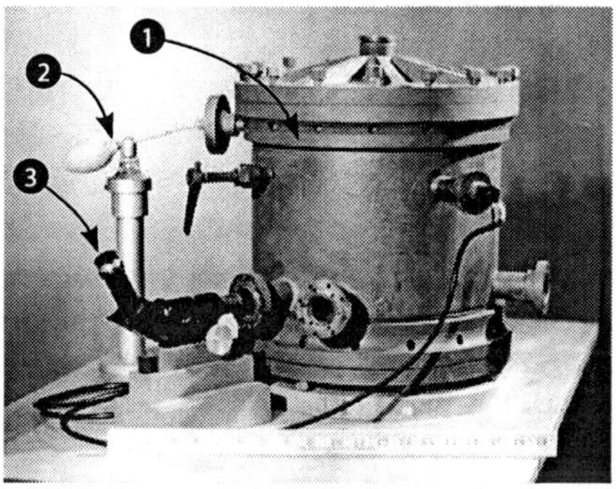

The process of squirting in the oil droplets with the atomizer sprayer caused some of the droplets to acquire a charge of static electricity. This means the droplets had excess electrons on them and carried a net negative charge. They picked up these extra electrons by friction as the droplets squirted through the rubber sprayer tube. As Millikan looked at an oil droplet through the eyepiece and adjusted the voltage between the plates, he could make the charged oil droplet hover when the voltage was just right. Millikan took into account the weight of the droplets and the viscosity of the air as the droplets fell and was able to determine that every droplet had a charge on it that was a multiple of 1.6×10^{-19} C. From this he deduced that this must be the charge on a single electron, which it is. Millikan won the Nobel Prize in Physics in 1923 for this work.

Figure 2-8. Apparatus for the oil-drop experiment.

Figure 2-9. New Zealander and physicist Ernest Rutherford.

The last famous experiment in this basic history of atomic models was initiated in 1909 by one of Thomson's students at Cambridge, New Zealander Lord Ernest Rutherford (Figure 2-9). Rutherford was already famous when this experiment occurred, having just won the Nobel Prize in Chemistry the previous year. Rutherford's *gold*

13

Chapter 2

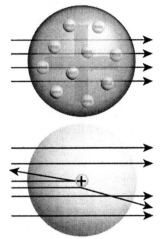

Figure 2-10. Alpha-particle pathways through the gold atoms expected from Thomson's plum pudding model (top) and the pathways Rutherford observed (bottom).

foil experiment resulted in the discovery of the atomic nucleus. To understand this experiment you need to know that an alpha particle, or α-particle, using the Greek letter alpha, is a particle composed of two protons and two neutrons. Alpha particles are naturally emitted by some radioactive materials in a process called *nuclear decay*.

Rutherford created a beam of α-particles by placing some radioactive material (radium bromide) inside a lead box with a hole in one end. The α-particles from the decaying radium atoms streamed out of the hole at very high speed (15,000,000 m/s!). Rutherford aimed the α-particles at an extremely thin sheet of gold foil that was only a few hundred atoms thick. Surrounding the gold foil was a ring-shaped screen coated with a material that would glow when hit by an α-particle. Rutherford could then determine where the α-particles went after encountering the gold foil.

When Rutherford began taking data, he had Thomson's plum pudding model in mind and was expecting results consistent with that atomic model. This scenario is depicted in the upper part of Figure 2-10. The atom's positive charge is spread throughout the atom and the negative electrons are embedded in the positive material. Rutherford expected the massive and positively charged α-particles to blow right through the gold foil.

What Rutherford found was astonishing. Most of the α-particles passed straight through the foil and struck the screen on the other side, just as Rutherford expected. However, occasionally an α-particle (one particle out of every several thousand) would deflect with a small angle. And sometimes the deflected particles would bounce almost straight back. This situation is depicted in the lower part of Figure 2-10.

The astonished Rutherford commented that it was like firing a huge artillery shell into a piece of tissue paper and having it bounce back and hit you! Rutherford's work led to his new proposal in 1911 for a model of the atom. Rutherford's model included the key points listed in Table 2-2.

In 1917, Rutherford became the first to "split the atom." In this experiment he used α-particles again, this time striking nitrogen atoms. His work led to the discovery of the positively-charged particles in the atomic nucleus, which he named protons.

It took another twenty years before James Chadwick (Figure 2-11), another Englishman, discovered the neutron. Before World War I, Chadwick studied under Rutherford at Cambridge. Then the war began. Not only did the war interrupt the progress of the research in general, but Chadwick was a prisoner of war in Germany. Working back in England after the war he discovered the neutron in 1932 and received the Nobel Prize in Physics for his discovery in 1935.

1. The positive charge in atoms is concentrated in a tiny region in the center of the atom, which Rutherford called the nucleus.
2. Atoms are mostly empty space.
3. The electrons, which contain the atoms' negative charge, are outside the nucleus.

Table 2-2. The main ideas in Ernest Rutherford's 1911 atomic model.

Atoms and Atomic Theory

Chadwick's discovery of the neutron enabled physicists to fill in a lot of blanks in their understanding of the basic structure of atoms. But years before Chadwick made his discovery, Rutherford's atomic model was already being taken to another level through the work of Niels Bohr. We will explore Bohr's atomic model, and the quantum model to which it led, in the main text.

Questions

1. Write a paragraph describing the atom. In your paragraph, make use of the following terms: proton, neutron, nucleon, nucleus, electron, and orbital.

2. Referring to the Periodic Table of the Elements on the inside back cover of *Chemistry for Accelerated Students*, write the chemical symbol and ionic charge for each of the elements listed below. Example: calcium: Ca^{2+} The answers are below, but cover them up until you are ready to check your work.

Figure 2-11. English physicist James Chadwick.

   a. sodium   b. bromine   c. barium   d. nitrogen
   e. oxygen   f. hydrogen   g. scandium   h. argon

3. Write paragraphs describing the experiments performed by J.J. Thomson, Robert Millikan, and Ernest Rutherford.

4. Compare and contrast the main points or features in the atomic models proposed by John Dalton, J.J. Thomson, and Ernest Rutherford.

5. Explain why Ernest Rutherford found the reflection of alpha-particles off gold foil so astonishing.

6. Explain why the first two points in Dalton's atomic model are now considered partially correct.

7. Identify the discoverer and year for each of the following: the nucleus, the proton, the neutron, and the electron.

Answers to question 2:

   a. Na^+ b. Br^- c. Ba^{2+} d. N^{3-}
   e. O^{2-} f. H^+ g. Sc^{3+} h. (does not ionize)

Chapter 3   Substances and Their Properties

Types of Substances

A *substance* is anything that contains matter. There are several major classifications of substances, but as shown in Figure 3-1, they all fall into two major categories, *pure substances* and *mixtures*.

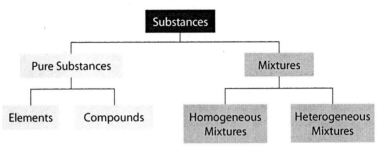

Figure 3-1. Classifications of substances.

Pure Substances

There are two kinds of pure substances, *elements* and *compounds*. We will discuss elements first. The Periodic Table of the Elements lists all of the known elements. This famous table plays a major role in the study of chemistry. We will dive into the periodic table in detail in the main text. I bring up the table here to assist in our discussion of elements, and you may wish to refer to the periodic table inside the rear cover of the main text during the following discussion.

- *Elements*

    The characteristic that defines each element in the periodic table is the number of protons the element has in each of its atoms, a number called the *atomic number*. The elements are ordered in the periodic table by atomic number. For example, carbon is element number six in the periodic table. This means that an atom of carbon has six protons—*all* carbon atoms have six protons. If an atom does not have six protons, it is not a carbon atom, and if an atom does have six protons, it is a carbon atom. An element is therefore a type of atom, classified according to the number of protons the atom has. A lump of elemental carbon—which could be graphite, diamond, coal, or several other varieties of pure carbon—is any lump of atoms that contain only six protons apiece. Oxygen is another example of an element. Pure oxygen is a gas (ordinarily) that contains only atoms with eight protons each, because oxygen is element number eight. Other examples of elements you have heard of are iron, gold, silver, neon, copper, nitrogen, lead, and many others.

    For every element, there is a *chemical symbol* that is used in the periodic table and in the chemical formulas for compounds. For some elements, a single uppercase letter is used, such as N for nitrogen and C for carbon. For other elements, an upper case letter is followed by one lower case letter, such as Na for sodium and Mg for magnesium. The

Substances and Their Properties

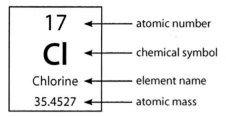

Figure 3-2. The basic information in each cell of the PTE.

three-letter symbols for elements 113, 115, 117, and 118 are placeholders until official names and two-letter symbols are selected by the appropriate governing officials.

Some of the chemical symbols are based on the Latin names of elements, such as Ag for silver, from its Latin name *argentum*. Other examples are Au for gold, from the Latin *aurum*, and Pb for lead, from the Latin *plumbum* (everyone's favorite Latin name).

The most common representations of the periodic table show four pieces of information for each element, indicated in Figure 3-2. At the top of the cell is the atomic number, symbolically represented as Z. Again, this number indicates the number of protons in each atom of the element, and is the number used to order the elements in the periodic table. Below the atomic number are the element's chemical symbol and name. At the bottom of the cell is the atomic mass. We will address the atomic mass in detail in the main text.

- *Compounds*

    As the name implies, a compound is formed when two or more different elements are chemically bonded together. This bonding is always the result of a chemical reaction. A chemical reaction is any process in which connecting bonds between atoms are formed or broken. Once bonded together chemically, the elements in a compound can only be separated by chemical means. In other words, it takes a different chemical reaction to break atoms apart.

    The elements or compounds that go into a chemical reaction are called the *reactants*. The elements or compounds formed by the reaction are called the *products*. The physical and chemical properties of a compound are completely different from the properties of any of the elements in the compound. Consider oxygen, hydrogen, and water. Hydrogen and oxygen react explosively to form water according to this *chemical equation*:

    $$2H_2 + O_2 \rightarrow 2H_2O$$

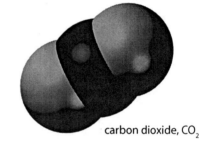

carbon dioxide, CO_2

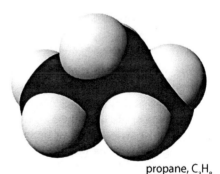

propane, C_3H_8

ozone, O_3

Figure 3-3. Common molecular substances.

Chapter 3

Oxygen is an invisible gas that we breathe in the air and that supports combustion. Hydrogen is an invisible, flammable gas. Water is composed of oxygen atoms bonded to hydrogen atoms, but one cannot breathe water, nor does water combust or support combustion. Or consider the sodium and chlorine in sodium chloride. These elements combine according to this chemical equation:

Na + Cl → NaCl

We all need salt in our diets, and we find it tasty. But both sodium and chlorine, the two elements of which sodium chloride is composed, are deadly dangerous in their pure, elemental forms. Sodium is a shiny, peach-colored metal that slices just like cheddar cheese. Chlorine is a poisonous, greenish gas. As you can see, the properties of these substances are nothing at all like the properties of the salt they form when they react together. We will discuss physical and chemical properties in more detail a bit later.

When atoms bond together to form a compound, the atoms in the compound can be arranged in two different basic types of structures. In many cases, the atoms join together to form molecules. In every molecule of a given substance, the atoms bond together in the same whole-number ratio. A few well-known molecular substances are carbon dioxide (CO_2), propane (C_3H_8), and ozone (O_3), all represented by space-filling models in Figure 3-3. The standard color coding used in computer models is white for hydrogen, black or charcoal gray for carbon, and red for oxygen. In the black and white images in Figure 3-3, gray is used in place of red to represent oxygen atoms.

It is important to note that in any chemical formula for a molecular substance, the formula indicates the number of each type of atom in the molecule. Propane, C_3H_8, has three carbon atoms and eight hydrogen atoms in each molecule. The subscripts are only shown when the quantity of atoms of an element is greater than one. The formula for sucrose—table sugar—is $C_{12}H_{22}O_{11}$. Each molecule of sucrose contains 12 carbon atoms, 22 hydrogen atoms, and 11 oxygen atoms.

The other common way atoms combine is by forming a continuous geometric arrangement. These compounds are called *crystals*, and the structure the atoms in the compound make when they join together is called a *crystal lattice*. The number of different arrangements atoms can make in a lattice is endless, and these arrangements are responsible for many of the unusual properties crystals possess. But what all lattices have in common is the regular arrangement of the atoms into repeating, geometrical patterns. A space-filling model of the very simple crystal structure for sodium chloride, NaCl (table salt), is shown in Figure 3-4. The formula tells us that the ratio of sodium atoms to chlorine atoms in the crystal is 1 : 1. The model shows that the atoms are bonded together in a simple alternating arrangement.

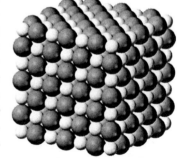

Figure 3-4. A space-filling model of the sodium chloride lattice structure. Sodium atoms are light gray and chlorine atoms are dark gray.

We have seen several space-filling models so far. Another common type of computer model is the ball-and-stick model. Figure 3-5 shows a ball-and-stick model of the somewhat more complex crystal structure of copper (II) chloride, $CuCl_2$. As the formula indicates, in the lattice structure there are two chlorine atoms for each copper atom.

Mixtures

So far we have been discussing one major category of substances—pure substances. Elements and compounds are pure substances. The other major category is *mixtures*. Any time substances are mixed together without a chemical reaction occur-

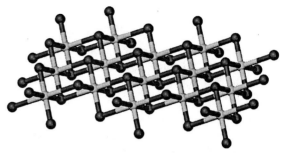

Figure 3-5. A ball-and-stick model of the copper (II) chloride crystal structure. Copper atoms are light gray and chlorine atoms are dark gray.

ring a mixture is formed. Remember—if a chemical reaction occurs, compounds are formed, not mixtures. If you toss vegetables in a salad, you've made a mixture. If you put sugar in your tea or milk in your coffee, you've made a mixture. If you mix up a batch of chocolate chip cookie dough, a bowl of party mix, or the batter for a vanilla cake, you've made a mixture.

In contrast to compounds, when a mixture is formed the individual substances in the mixture retain their physical and chemical properties. If you mix salt in water, you've made a mixture. The salt is still there and tastes salty. The water is still there too, and tastes watery. Also in contrast to compounds, the substances in a mixture can be separated by physical means such as filtering, boiling, freezing, or settling.

Again, in compounds, the original properties of the elements in the compound are chemically changed into the properties of the new chemical substance that is formed by means of a chemical reaction. Further, the elements in a compound cannot be separated by physical means. They can only be separated by the same means that brought them together in the first place—a chemical reaction. The distinguishing features of mixtures and compounds are summarized in Table 3-1. There are two classes of mixtures—*homogeneous mixtures* and *heterogeneous mixtures*. We will examine these next.

Mixtures	Compounds
• Formed when substances combine without a chemical reaction occurring.	• Elements combine chemically to form a new substance—a compound.
• The individual substances in the mixture retain their physical and chemical properties.	• The physical and chemical properties of the compound are completely different from those of the reactants that formed the compound.
• The substances in mixtures can be separated by physical means such as filtering, boiling, freezing, or settling.	• The elements in a compound can only be separated by chemical means.

Table 3-1. Summary of the distinctions between mixtures and compounds.

Chapter 3

- *Homogeneous Mixtures*

  Homogeneous mixtures have uniform composition down to, but not including, the groups of atoms at the atomic level. The individual particles of the different substances in a homogeneous mixture cannot be seen with the eye, not even with the most powerful microscope. Particles at the atomic level—atoms, molecules, ions—are too small to reflect visible light, and so cannot be seen in the ordinary way, regardless of the magnification.

  The implication of this definition is that homogeneous mixtures are identical with *solutions*,[1] mixtures in which one pure substance (or more than one) is dissolved in another pure substance. Solutions are of major importance in chemistry because many chemical reactions take place in aqueous solution (a solution in water, as you recall). We will devote an entire chapter to solutions in the main text.

- *Heterogeneous Mixtures*

  In contrast to homogeneous mixtures, in a heterogeneous mixture there are lumps of different substances mixed together. You might be able to see the lumps with the naked eye, as with the mixture of spices in meat seasoning. Or the different substances may be visible only under a microscope, such as microscopic particulates in well water. Either way, if the different substances can be seen, the mixture is heterogeneous. In addition to the examples of salads and so on I mentioned above, there are two classes of heterogeneous mixtures that we encounter every day: *suspensions* and *colloidal dispersions*. A *suspension* is formed when particles of size approximately 1 micrometer (1 μm) or larger are dispersed in a fluid (liquid or gas) medium. Particles this large will not remain in suspension indefinitely; they will eventually settle out due to gravity. An example of a suspension is muddy water. *Colloidal dispersions*, usually referred to simply as *colloids*, are formed when microscopic particles, ranging in size from 1 to 1,000 nm, are dispersed throughout a dispersing medium. Unlike the particles in suspensions, these particles will not settle out. Forces from *Brownian motion* (collisions from other molecules of air, water, etc.) keep the particles dispersed in the dispersing medium. Examples of colloids are fog and gelatin.

  Further discussion of suspensions and colloids awaits us in the main text.

Physical and Chemical Properties

All substances have certain properties. We divide up the different properties substances can possess into two broad classes. Some properties have to do with the physical characteristics of the substance, such as color, shape, size, phase, boiling point, texture, thermal conductivity, electrical conductivity, opacity, and density. These properties are called *physical properties*. Below is a list of example statements about the physical properties of substances. Consider how each one relates to the definition of physical properties just given.

- Iron is gray in color.
- Copper has a high electrical conductivity.

---

[1] In the past, sources have differed in their definitions of homogeneous mixtures. As a result, my own definitions in previously written texts have differed from the definition given here. More recently, the definition given here (equating the terms homogeneous mixture and solution) seems to be preferred, although some sources still differ.

Substances and Their Properties

- Mica is shiny.
- Glass is smooth, but has sharp edges.
- Aluminum has a high thermal conductivity.
- Ethyl alcohol is transparent and colorless.
- Chlorine is greenish-yellow gas at atmospheric pressure and room temperature.
- Helium is a gas at atmospheric pressure and room temperature.
- At standard pressure, water freezes at 0°C.
- Milk is opaque.
- Oil is slippery.
- Clay brick is ochre in color.
- At 4.0°C, the density of water is 1.0 g/cm^3.
- Gold is malleable and ductile.
- Cast iron is not malleable.
- Glass is not malleable.
- Modeling clay is malleable, but not ductile.
- Jello is not ductile.

You may have noticed a couple of unfamiliar terms in the above list. The terms *malleable* and *ductile* are used to describe two important properties possessed by many metals. A substance is malleable if it can be hammered into different shapes, or hammered flat into sheets. A substance is ductile if it can be *drawn* into a wire. Wire drawing is a process of making wire by pulling the metal through a small hole in a metal block called a *die*. Usually the metal is already formed into a wire of larger diameter. The end of this larger diameter wire is hammered down or filed to get it through the hole in the die, and then a machine pulls the wire through the die to make the new, smaller diameter wire. Substances that can be drawn through a die like this without simply snapping are said to be ductile.

Notice from the above examples that a good student of science needs to be careful when describing physical properties. We need to make sure our statements are accurate in cases where temperature or pressure affect the property in question. For example, it is inaccurate to say that H_2O is a liquid. A more accurate statement would be to say that one of the physical properties of water is that it is a liquid at temperatures between 0°C and 100°C. An even more accurate statement would be to specify that the preceding sentence is correct at atmospheric pressure, because at other pressures the boiling and freezing points of water are different.

The second broad class of properties has to do with the kinds of chemical bonds a substance will form, that is, the chemical reactions a substance will or will not participate in. These properties are called *chemical properties*. We have not yet studied chemical reactions, so you may not know that much about them. However, there are two common chemical reactions that you are quite familiar with—burning and rusting. Both of these are chemical reactions in which a substance combines with oxygen, and both are examples of a type of reaction called *oxidation*. Fiery explosions are simply combustions that happen very rapidly. But whether the combustion happens slowly, as with a log on a fire, or rapidly, as with a firecracker, combustion is a chemical reaction with oxygen. Substances that will react with oxygen in this way are said to be *flammable* or *combustible*. (Oddly, *inflammable* means the same thing!)

Iron oxidizes to form compounds known as *oxides*. There are several different forms of iron oxide, colored red, yellow, brown, and black. Other metals oxidize as well. When copper oxidizes it can form two different oxides, one red and one black. This is why copper objects exposed to the air will turn dark brown or black. (Over a longer period of time the copper oxide forms other compounds, such as copper carbonate, which give the copper its pretty blue-green color. The Statue of Liberty is made of copper, and has been there for a long time. It is essentially covered with a layer of copper carbonate and other copper compounds.) Aluminum also oxidizes. Aluminum oxide is dark gray, and anyone who has done a lot of hand work with aluminum parts will have noticed his or her hands blackened by the particles of aluminum oxide building up.

Here are some examples of how one would describe chemical properties of substances:

- Hydrogen is combustible.
- Aluminum oxidizes to form aluminum oxide.
- Water is not flammable.
- Platinum does not oxidize. (This is why it is so valuable. It stays shiny as a pure element.)
- Baking soda (sodium bicarbonate, $NaHCO_3$) reacts with vinegar (acetic acid, CH_3COOH).
- Iron oxidizes to form iron oxide, or rust.
- Sodium reacts violently with water.
- Hydrogen reacts with a number of different polyatomic ions to form acids.
- Dynamite is explosive.
- Sodium hydroxide reacts with aluminum.
- Sulfuric acid reacts with many metals.

We will be discussing the physical and chemical properties of substances even more in the main text.

The two broad classes of properties we have been discussing, physical properties and chemical properties, are related to two broad classes of changes that substances can undergo. If a substance experiences a change with respect to one of its physical properties, we call this a *physical change*. When a physical change occurs, the substance is still the same substance, it just looks different. If a chemical reaction occurs to a substance, this is called a *chemical change*. Chemical properties basically describe the kinds of chemical changes a substance can undergo. When a chemical change occurs, the original substances that went into the reaction—the reactants—are converted into new substances—the products—with totally different physical and chemical properties. When asked to describe a given change as physical or chemical, ask yourself if the substance is still the same substance, or if it has actually gone through a chemical reaction and has become a different substance.

Table 3-2 contains a number of examples of physical and chemical changes, with comments explaining why the type of change is physical or chemical.

Density

Density is a physical property of substances. Density is a measure of how much matter is packed into a given volume for different substances. No doubt you are already familiar with the concept of density. You know that if you hold equally sized balloons in each hand, one filled with water and one filled with air, the water balloon will weigh more because water is denser than air. You know that to have equal weights of sand and Styrofoam packing peanuts the volume of the packing material will be much larger because the packing material is much less dense. And you probably also know that objects less dense than water will float, while objects denser than water will sink.

Process	Change	Comments
glass breaking	physical	The broken glass is still glass, it just changed shape.
firecracker exploding	chemical	This explosion is a combustion. All combustions are chemical reactions. The substances in the firecracker have reacted to form new substances such as ash and various gases.
mercury boiling	physical	The mercury is still mercury, it has just changed from the liquid phase to the vapor phase.
copper turning dark brown or black	chemical	This occurs because the copper is oxidizing and forming copper oxide, a new substance. This is a chemical reaction.
iron pipes corroding	chemical	Corrosion is a chemical reaction. In this case, the iron reacts with the substances surrounding the pipes to form a new substance.
water evaporating	physical	The substance is still H_2O, it has simply changed phase from liquid to vapor.
mixing cake batter	physical	The eggs and flour and so on have formed a mixture, but no chemical change (reaction) has occurred.
baking cookies	chemical	The heat caused a chemical reaction to occur in the dough. The substance is no longer cookie dough. It is cookie.
spilled pancake batter drying out	physical	No chemical reaction occurred. Dried batter is still batter. (If you want a pancake you have to cook it, which would be a chemical change.)
molten lead hardening	physical	The lead is still lead, it just changed phase from liquid to solid.
balloon popping	physical	The balloon material is still the same material, it is just in shreds now. The air inside the balloon is at a lower pressure and is not contained in the balloon any longer, but it is still air.

Table 3-2. Examples of physical and chemical changes.

The equation for density is

$$\rho = \frac{m}{V}$$

where the Greek letter ρ (spelled rho and pronounced "row," which rhymes with snow) is the density in kg/m³, m is the mass in kg, and V is the volume in m³. These are the variables and units in the MKS unit system. However, since laboratory work typically involves only small

Chapter 3

quantities of substances, it is more common in chemistry for densities to be expressed in g/cm³ (for solids) or g/mL (for liquids). In the examples that follow I will illustrate the use of g/cm³ and kg/m³. Since 1 mL = 1 cm³, calculations solving for densities in g/mL are essential the same as those solving for densities in g/cm³. If all you are doing is using the density equation, then any of these units of measure is fine. One final item for you to note is that the density of water at room temperature is

$$\rho_w = 0.998 \, \frac{g}{cm^3} \quad (22.0°C)$$

This value comes up all the time in chemistry and physics, and it is well to commit it to memory. We will now work through two examples of calculations involving density.

The density of germanium is 5.323 g/cm³. A small sample of germanium has a mass of 17.615 g. Determine the volume of this sample.

Begin by writing the given information.

$$\rho = 5.323 \, \frac{g}{cm^3}$$

$$m = 17.615 \, g$$

$$V = ?$$

Now write the density equation and solve for the volume.

$$\rho = \frac{m}{V}$$

$$\rho \cdot V = m$$

$$V = \frac{m}{\rho}$$

Next, insert the values and compute the result.

$$V = \frac{m}{\rho} = \frac{17.615 \, g}{5.323 \, \frac{g}{cm^3}} = 3.309 \, cm^3$$

This value has four significant digits, as it should based on the given information.

Determine the density of a block of plastic that has a mass of 1,860 g and dimensions 4.0 × 2.5 in × 9.50 in. State your result in kg/m³.

To solve this problem, we will use the given dimensions to calculate the volume of the block. Then we will use the density equation to calculate the density.

24

Always begin your problem solutions by writing down the given information and performing any necessary unit conversions. Since the units of measure required for the result are kg/m³, we will convert the mass to kilograms and the lengths to meters. When solving problems requiring unit conversions like this, write down the given information on separate lines down the left side of your page. Then perform the unit conversions by multiplying the conversion factors out to the right.

From the given information, our result must have two significant digits. This means we must perform the unit conversions and volume calculation with three significant digits (one more than we need), and round to two digits when we get our final result.

$$m = 1860 \text{ g} \cdot \frac{1 \text{ kg}}{1000 \text{ g}} = 1.86 \text{ kg}$$

$$l = 4.0 \text{ in} \cdot \frac{2.54 \text{ cm}}{1 \text{ in}} \cdot \frac{1 \text{ m}}{100 \text{ cm}} = 0.102 \text{ m}$$

$$w = 2.5 \text{ in} \cdot \frac{2.54 \text{ cm}}{1 \text{ in}} \cdot \frac{1 \text{ m}}{100 \text{ cm}} = 0.0635 \text{ m}$$

$$h = 9.5 \text{ in} \cdot \frac{2.54 \text{ cm}}{1 \text{ in}} \cdot \frac{1 \text{ m}}{100 \text{ cm}} = 0.241 \text{ m}$$

Now that the units are squared away, let's determine the volume of the block.

$$V = l \cdot w \cdot h = 0.102 \text{ m} \cdot 0.0635 \text{ m} \cdot 0.241 \text{ m} = 0.00156 \text{ m}^3$$

Finally, using the density equation the density is

$$\rho = \frac{m}{V} = \frac{1.86 \text{ kg}}{0.00156 \text{ m}^3} = 1192 \frac{\text{kg}}{\text{m}^3}$$

Rounding to two significant digits we have

$$1200 \frac{\text{kg}}{\text{m}^3}$$

Questions

1. Write paragraphs distinguishing between these pairs of terms:
    a. compounds and elements
    b. mixtures and compounds
    c. heterogeneous mixtures and homogeneous mixtures
    d. suspensions and colloids
2. Classify each of the following as element, compound, homogeneous mixture, or heterogeneous mixture.

Chapter 3

a. water	b. cesium chloride	c. pond water	d. methane
e. a soft drink	f. nitric acid	g. black coffee	h. argon
i. air	j. hydrogen nitrate	k. exhaust fumes	l. quartz
m. brass	n. hydrogen gas	o. hydrogen cyanide	p. mouthwash
q. platinum	r. dirt	s. radon	t. a smoothie

3. Explain why salt water and sugar water are homogeneous mixtures while automotive paint, which contains invisible particulates, is not.

4. Write a paragraph describing the two basic types of structures atoms can take when bonding together.

5. Select three pure substances not mentioned in the chapter. For each substance, list at least eight physical properties and three chemical properties.

6. Identify each of the following as a physical change or a chemical change. For each, explain your choice.

   a. an avalanche
   b. a cigar burning
   c. spilling a glass of milk
   d. digesting your food
   e. swatting a fly
   f. stirring cream into coffee
   g. firing a pop gun
   h. firing a real gun
   i. boiling mercury
   j. welding steel
   k. filling a helium balloon
   l. allowing molten iron to harden
   m. frying chicken
   n. snow melting
   o. a car exhaust pipe rusting
   p. paint "drying"
   q. wood rotting
   r. a ball rolling down a hill

7. What is the density of carbon dioxide gas if 0.196 g of the gas occupies a volume of 100.1 mL? *(Ans: 0.00196 g/mL)*

8. Oil floats because its density is less than that of water. Determine the volume of 550 g of a particular oil with a density of 955 kg/m^3. State your answer in mL. *(Ans: 580 mL)*

9. A factory orders 15.7 kg of germanium. The density of germanium is 5.32 g/cm^3. Calculate the volume of this material, and state your answer both in m^3 and cm^3. *(Ans: 0.00295 m^3, 2,950 cm^3)*

10. A graduated cylinder contains 23.35 mL of water. An irregularly shaped stone is placed into the cylinder, raising the volume to 27.79 mL. If the mass of the stone is 32.1 g, what is the density of the stone? *(Ans: 7.23 g/cm^3)*

11. A standard 55-gallon drum is 34.5 inches tall and 24 inches in diameter. Consider a 55-gallon drum filled with kerosene. Using the dimensions in inches to calculate the volume, determine the mass of kerosene that would fill this drum, given that the density of kerosene is 810 kg/m^3. *(Ans: 210 kg)*

12. Iron has a density of 7,830 kg/m^3. An iron block is 2.1 cm by 3.5 cm at the base and has a mass of 94.5 g. How tall is the block? *(Ans: 1.6 cm)*

Image Credits

2, 6. John D. Mays. 9. Vlada Marinković, licensed under CC-BY-SA-3.0. 10. Benjah-bmm27, public domain. 11. Arthur Shuster & Arthur E. Shipley: Britain's Heritage of Science. London, 1917. Based on a painting by R.R. Faulkner, public domain. 12. Thomson: GWS - The Great War: The Standard History of the All Europe Conflict (volume four) edited by H. W. Wilson and J. A. Hammerton (Amalgamated Press, London 1915); CRT: Mrjohncummings, licensed under CC-BY-SA-2.0. model: John D. Mays. 13. Millikan: The Electron: Its Isolation and Measurements and the Determination of Some of its Properties, Robert Andrews Millikan, 1917, public domain; Millikan's apparatus: Public domain; Rutherford: Public domain. 14. Fastfission, public domain. 15. Bortzells Esselte, Nobel Foundation, public domain. 16. John D. Mays. 17. carbon dioxide: Jynto, public domain; propane and ozone: Benjah-bmm27, public domain. 18. Benjah-bmm27, public domain. 19. Benjah-bmm27, public domain.